The Truth About Your Food

Be Smart About GMOs

By Rachael Morlock

Cavendish Square

Published in 2023 by Cavendish Square Publishing, LLC
2544 Clinton Street, Buffalo, NY 14224

Website: cavendishsq.com

Library of Congress Cataloging-in-Publication Data

Names: Morlock, Rachael, author.
Title: Be smart about GMOs / Rachael Morlock.
Other titles: Be smart about genetically modified organisms
Description: Buffalo, New York : Cavendish Square Publishing, [2023] |
Series: The truth about your food | Includes index.
Identifiers: LCCN 2022023413 | ISBN 9781502665799 (library binding) | ISBN 9781502665782 (paperback) | ISBN 9781502665805 (ebook)
Subjects: LCSH: Transgenic plants. | Crops–Genetic engineering. | Genetically modified foods.
Classification: LCC SB123.57 .M67 2023 | DDC 631.5/233–dc23
LC record available at https://lccn.loc.gov/2022023413

Editor: Rachael Morlock
Copyeditor: Shannon Harts
Designer: Deanna Paternostro

The photographs in this book are used by permission and through the courtesy of: Cover (main) margouillat photo/Shutterstock.com; cover (DNA graphic), back cover, pp. 3, 4, 7, 10, 15, 23, 27, 30, 35, 36, 42-43, 44-45, 46-47, 48 Omelchenko/Shutterstock.com; p. 5 Jon Rehg/Shutterstock.com; p. 6 PDH/Wikimedia Commons; p. 9 Volodymyr Dvornyk/Shutterstock.com; p. 10 Daniel Beckemeier/Shutterstock.com; p. 12 Budimir Jevtic/Shutterstock.com; p. 14 GOLFX/Shutterstock.com; pp. 16-17 Sourabh_Vyas/Shutterstock.com; p. 19 jadimages/Shutterstock.com; p. 22 Merrilee Holmes/Shutterstock.com; p. 23 inewsfoto/Shutterstock.com; p. 24 uladzimir zgurski/Shutterstock.com; p. 26 Debby Lowe/Shutterstock.com; p. 28 vincentchuls/Shutterstock.com; p. 30 Cindy Creighton/Shutterstock.com; p. 31 Fotokostic/Shutterstock.com; p. 33 LightField Studios/Shutterstock.com; p. 34 Peter Turner Photography/Shutterstock.com; p. 36 Tada Images/Shutterstock.com; p. 38 Sherry Saye/Shutterstock.com; p. 40 Bogdan Wankowicz/Shutterstock.com.

CPSIA compliance information: Batch #CW23CSQ: For further information, contact Cavendish Square Publishing LLC at 1-877-980-4450.

Printed in the United States of America

Contents

Introduction

A simple trip to the grocery store can involve a bewildering array of choices. Eye-catching packaging, health claims, and special labels convince buyers to choose one item over another. Some foods are labeled non-GMO, while others have a bioengineered label. What do these labels really mean?

GMOs, or genetically modified organisms, have been a hot topic as scientists continue to discover ways of changing, or genetically modifying, the DNA of plants and animals, which carries their genetic information. This technology is used to make food easier to produce or more desirable in some way. The process results in foods that are called genetically modified (GM), **genetically engineered** (GE), or bioengineered.

In the United States, three government organizations work together to regulate GMOs—the U.S. Department of Agriculture (USDA), the Environmental Protection Agency (EPA), and the Food and Drug Administration (FDA). With their approval, the first GE crops were marketed to Americans in 1994.

The major GE crops today are corn, soybeans, canola, cotton, alfalfa, and sugar beets. These crops are mainly used to feed food-producing animals, such as cows and chickens. However, GE crops do enter the grocery store, often in the form of packaged foods. **Processed food** ingredients like corn starch, corn syrup, soybean oil, canola oil, and granulated sugar are mostly made from plants that are grown from GE seeds. For foods in the produce aisle, genetic engineering has been used to develop new forms of potatoes, tomatoes, papayas, summer squash, and apples.

Animals can be genetically engineered as well, and a fast-growing type of salmon was the first GE animal product to be approved for sale in the United States.

Corn is one of the most common GMOs. More than 90 percent of the corn grown in the United States is genetically engineered.

Many people feel strongly about GMOs. Some think GMOs can solve problems like world hunger by increasing food production, boosting the nutritional value of foods, and decreasing the effects of global climate change on agriculture. Others believe that GMOs pose risks to our health and the environment.

A lot of opinions about GMOs are influenced by emotions or politics. There's room for debate because GMOs are not all bad or all good. Genetic modification is a tool that can be used to improve food, but it can also cause harm if not used carefully. While sorting through these differing viewpoints, it can be difficult to find reliable facts about GMOs. However, learning about the benefits and risks of GMOs can help you make mindful decisions about what you eat and why.

A bioengineered tomato called Endless Summer was one of the first GM tomatoes developed in the 1990s.

Chapter One

GMOs Explained

All organisms have specific traits, or qualities, that are determined by the genes in their DNA. Genetic modification deals with changing the genes of an organism. Genetic engineering refers more specifically to the process of adding a gene from another organism to introduce a new trait. Most of the GMOs in our food system are engineered with genes that make them resist **pesticides**, pests, bacteria, or viruses. Others are engineered to survive drought or to contain **vitamins** they don't naturally contain. More recently, apples and potatoes have been modified with genes that prevent browning when they're cut—a change that is hoped to reduce food waste. There's also plenty of room for experimentation in the ways genetic engineering can tweak the flavors of fruits and vegetables to make them more appealing. The process of cultivating plants with beneficial traits is as old as farming—about 10,000 years old. However, the techniques used to make GMOs today have been developed only in the last 50 years.

The History of Genetic Modification

Humans have been modifying plants for thousands of years through **artificial selection**, also called selective breeding. To understand artificial selection, imagine a farmer who grows tomatoes. Among his tomato plants, one produces many more tomatoes than the others. Whatever makes that one plant more productive is encoded in its DNA. The farmer wants to grow as many tomatoes as possible, so the next season, he plants seeds only from that very productive tomato plant. The other tomato plants don't get to pass on their DNA to the next generation since their seeds aren't used. Instead, only the DNA of the plant chosen by the farmer is passed on. By growing plants that exhibit desirable qualities, farmers influence the genetic makeup of their plants. These desirable qualities are traits encoded in genes.

Starting in the 19th century, scientists began to understand how genes were inherited, or passed down. They made the connection between physical traits and DNA in the 20th century and realized that genes are encoded in DNA. In the 1950s, **biophysicists** uncovered the structure of DNA. As research continued, they learned how to recognize and separate individual genes. Once that was possible, specific genes could be inserted into the DNA of other organisms.

Transferring genes from one organism to another is how GMOs are usually created. Say you find an organism, like a bacterium, that can keep away or even kill a particular insect. This insect loves eating corn, and you want your corn to be safe from the insect, just like the bacteria.

You examine the bacteria's DNA, find the gene responsible for insect **resistance**, copy it, and then insert it into the DNA of one corn cell. Eventually, your one corn cell will divide into two, then

Gregor Mendel, who is depicted here, is considered the father of genetics. His experiments in breeding peas in the mid-1800s helped explain how traits are passed down.

GE Medicine

Early efforts in genetic engineering were applied to making medicines. A major breakthrough was the use of genetically engineered insulin to treat diabetes. Insulin is a hormone that the body needs for breaking down food and regulating blood sugar, but people with diabetes do not produce enough insulin on their own.

Before genetic engineering was understood, the only insulin for treating diabetes came from animals, especially cows and pigs. Doctors delivered the first dose of animal insulin to a patient with diabetes in 1922. The treatment was successful, but doctors found that insulin from animals could cause allergic reactions in humans.

The discovery of genetic engineering offered an opportunity to improve insulin treatments. Scientists found the genetic code that was responsible for the creation of insulin in humans. In 1978, they inserted the gene for insulin production into the DNA of bacteria called *E. coli*. Since genes are the code to create, or synthesize, proteins, the bacteria now had the tools to make insulin, which is a protein. By applying genetic engineering to medicine, scientists were able to make the first synthetic human insulin. Since then, genetic engineering has been an important tool for making new medicines and vaccines.

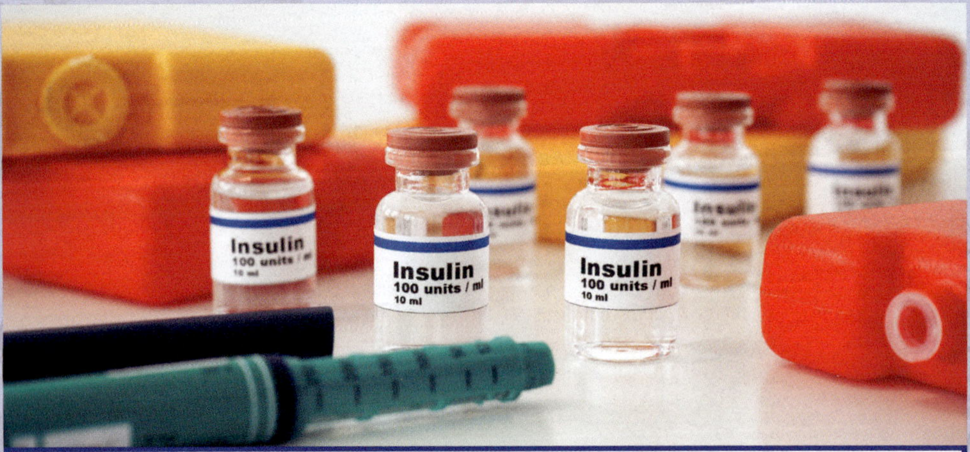

In 1982, insulin became the first product defined as a GMO to be approved by the FDA.

those cells will divide, and so on. Finally, you will have a plant that is resistant to the insect you targeted. An organism like this corn is called transgenic because the genes of another organism have been transferred to it.

GMO Timeline

1973: Herbert Boyer and Stanley Cohen create genetic engineering. They were the first to insert DNA from one bacterium into another.

1992: The FDA says GM food must meet the same safety requirements as non-GMOs.

1994: The first GE food—a tomato—is approved in the United States.

2003: The World Health Organization (WHO) and the Food and Agriculture Administration (FAO) of the United Nations create international rules and standards for GMOs.

2007: The Non-GMO Project is founded to certify foods free from GMOs.

2014: About 90 percent of the corn, cotton, and soybeans planted in the United States are GMOs.

2015: The FDA approves the first genetically engineered animal—GM salmon—as food.

2016: U.S. Congress passes the National Bioengineered Food Disclosure Law that requires labels on most foods containing GMOs.

2017: GM apples are sold in the United States.

2022: New labels and logos are applied to foods to mark them "Bioengineered" and "Made with Bioengineered Ingredients."

As new technologies have developed over time, genetic engineering has changed the way many foods are produced, regulated, and sold.

Good or Bad?

Genetic engineering is a more sophisticated way of doing what humans have been doing for thousands of years—making organisms, especially plants, inherit and exhibit desirable traits. Technological advances allow this process to unfold quickly and precisely instead of across generations. In the past, farmers who wanted a more productive crop would choose to use seeds from the plant that was most productive, though they didn't know what made it so. Now, scientists are able to identify the genes responsible for elements of productivity. The idea is the same, but the methods have changed.

GMO alfalfa is engineered to resist herbicides, or weed killers. Alfalfa is mainly used for feeding dairy cows.

Some people worry that changing the genetic makeup of food can have a negative impact on their health. GE seeds were introduced in the 1990s, and since then many studies have investigated their effects. Based on these studies, 90 percent of scientists agree that GE foods are safe to eat. The American Medical Association, the National Academy of Sciences, the British Royal Society, and the World Health Organization (WHO), among other scientific organizations, have stated that GE food is not more dangerous than non-GMO versions of the same food. However, the WHO and Royal Society do caution that the safety of GMOs should be tested and evaluated on a case-by-case basis. Testing helps ensure that GE foods will not cause allergic reactions in people who eat them.

Most of the GE foods made in the United States are fed to animals. In fact, billions of animals are fed primarily on GE ingredients. Studies show no differences in the health of animals that eat GE ingredients and those that do not.

Although GE foods may be safe to eat, the industrial farming methods that are used to grow them are cause for concern. Some people avoid GE foods because they are produced with toxic pesticides that pollute the environment and cause sickness in people. These farming practices mean that some GMOs can have negative indirect effects on our environment and health.

Food for Thought

1. What are the differences between artificial selection and genetic engineering?
2. Why do you think GE medicine is less debated than GE food?
3. What trusted sources do you use for information about food safety?

Cowpeas are a protein-rich food that can be ground into flour and cooked into meals. Increasing the insect resistance of cowpea plants can be achieved through genetic engineering.

Chapter Two

The Purpose of GMOs

The challenge of feeding the world is growing as the global population increases. The number of people living on Earth has increased from 1.65 billion at the beginning of the 20th century to more than 7.9 billion today. Of that number, a large proportion of people experience hunger. In 2020, 2.3 billion people around the world did not have access to enough food throughout the year.

Many scientists see GMOs as a way to feed people more efficiently. Although the world's population has increased, the planet and its resources haven't. GMOs can be part of the solution to growing a larger and more nutritious food supply with limited natural resources.

Global climate change also affects agricultural production. There's a greater need for crops that can withstand extreme weather, especially floods and droughts, while also resisting pests and diseases. GE solutions to world hunger aim to improve staple crops—the foods people rely on most in their diets. New innovations focus on vitamin-enriched rice, virus-resistant cassava, insect-resistant cowpeas, and blight-resistant potatoes.

However, the majority of GE crops in the United States feed animals instead of people. They are destined for animal feedlots where meat and dairy are produced. Most of the remaining GM products are ingredients in processed foods that have little nutritional value. GMOs can help solve world hunger, but only when used as part of a healthy food system.

The Green Revolution

Improved irrigation, better weather monitoring, **crop rotation**, synthetic **fertilizers**, and seed varieties that produce a high yield have all made producing a greater amount of food possible. This collection of technologies and methods was promoted by scientists and governments in a movement called the green revolution. Research began in the 1930s, and the movement expanded in the 1960s in an effort to prevent world hunger.

One method introduced by the green revolution was double-cropping—growing two crops per year on existing farmland. Irrigation projects, like dams, made double-cropping possible, so farmers didn't have to rely only on the rainy season. Another way to increase food production was using high-yield seed varieties, of wheat and rice in particular. The green revolution was a big success in Mexico, Pakistan, and India.

Prior to the green revolution in India, wheat was imported. The revolution introduced a variety of wheat developed through artificial selection. Most of the energy in this wheat was stored in the kernel and not in

Today, wheat is the main cereal crop grown by farmers in India.

the stalk, which is inedible. As a result, farmers could get more grain per acre, or more food on less land. With this innovation, India began exporting more wheat than it imported.

Something similar happened with rice. In 1966, the International Rice Research Institute (IRRI) used **crossbreeding** to develop a strain of rice called India Rice 8 (IR8). With this strain, rice yield went from roughly 1 ton (3 metric tons) to 3 tons (7 metric tons) per acre. Techniques from the green revolution doubled Asia's yields of rice and wheat from the 1960s to the 1990s.

Although the green revolution greatly improved the efficiency of food production, the seed varieties it promoted also relied heavily on synthetic fertilizers and pesticides to produce high yields. IR8, for example, requires nitrogen fertilizer to trigger its short growth duration and high yield. Both fertilizers and pesticides have many negative effects—they are harmful, and at times deadly, to animals, to humans, and to whole ecosystems, including our oceans.

Problems with Pesticides

Pesticides are chemicals that control the effects pests like insects, weeds, fungi, and rodents can have on plants. Pesticides either kill pests or repel them from the plant. However, sometimes the same results can be achieved through genetic engineering. One way of reducing the use of pesticides is to change the DNA of the plant itself. Inserting a gene that makes a plant produce proteins that are toxic to certain insects results in plants with an inbuilt defense. If plants provide their own resistance to pests, then fewer pesticides are needed to help them grow.

Some GE crops are engineered to be resistant to insects. They are created with genes from *Bacillus thuringiensis* (*Bt*), a natural bacterium that organic farmers

Roundup is the brand name of the herbicide (weed-killing pesticide) glyphosate that is used by industrial farmers and small-scale gardeners.

also use to treat crops. When genes from *Bt* are inserted in a plant's DNA, the plant is able to make proteins that are toxic to insects but not toxic to humans or other animals. *Bt* cotton and *Bt* corn are engineered in this way. *Bt* crops are designed to target certain insects so that pesticides aren't needed.

A second approach to genetically modifying plants is to make them herbicide-tolerant. Herbicides are chemical pesticides designed to kill unwanted weeds. Herbicide-tolerant (HT) crops are engineered to be resistant to the herbicide glyphosate, also called Roundup. Corn, soybean, and cotton crops that tolerate this herbicide are called "Roundup Ready." Herbicide-tolerance allows the crops to be sprayed with glyphosate to

Percent of Crops Grown with GE Seeds

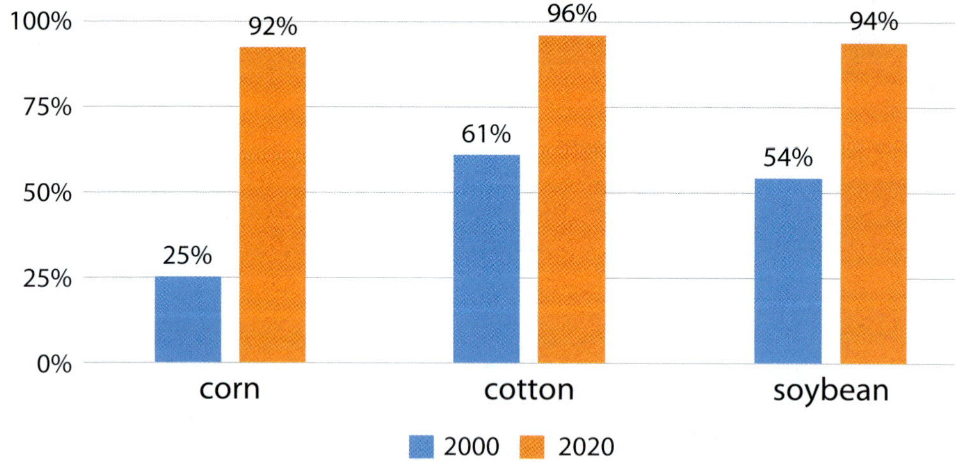

From the year 2000 to 2020, the percentage of corn, cotton, and soybean crops that are genetically modified has increased in the United States. Today, most of these crops are GMOs.

kill weeds around the crop without killing the crop itself. This GE development has helped glyphosate rise in popularity because farmers can use it to kill weeds without worrying that it will kill important crops.

When HT crops were first developed, glyphosate was less toxic than other herbicides in use. More recently, glyphosate has been linked to serious health risks for people in close contact with the chemical. Many farmworkers have developed the cancer non-Hodgkin's lymphoma after working with Roundup. Glyphosate also affects ecosystems by killing beneficial plants and polluting water systems. Despite the risks, the number of crops grown with herbicide-resistant seeds has increased in the United States over the past few decades.

Resilient and Nutritious

In addition to insect-resistant and herbicide-tolerant plants, genetic engineering can also create plants that are more **resilient**. One desirable quality for organisms is virus resistance. A plant virus has the ability to hurt plants and sometimes kill them. Papaya fruit in Hawaii, for example, were being killed by the papaya ringspot virus (PRSV) before genetic engineering stepped in. This virus was discovered in 1992 in Puna, Hawaii, where the majority of Hawaiian papaya was grown. Between 1992 and 1998, production fell from 53 million pounds (24 million kilograms) to 26 million pounds (11 million kg). The introduction of a virus-resistant papaya, called the Rainbow papaya, allowed production to climb back up to 40 million pounds (18 million kg) by 2001.

Today, our food supply is affected by climate change. The extreme weather resulting from climate change means there are more and longer droughts. This impacts crops since crops depend on rainfall. Some GE crops are specially engineered to

In the case of the genetically modified Rainbow papaya, the ringspot virus was introduced to the DNA of the fruit so that it would be immune to the virus.

survive drought. Crops that can endure flooding are another valuable product of genetic engineering. In Bangladesh, a GMO called scuba rice can survive underwater for up to 14 days. This is an important quality in Bangladesh and India, where farmers lose 4 million tons (3.6 million metric tons) of rice each year because of flooding.

Gene Editing

The process of changing DNA has become more advanced since genetic engineering was developed. In addition to adding genes from other organisms, it's now possible to remove genes from DNA. This is called gene editing, and technologies like CRISPR and TALEN make it possible. Think of these tools as tiny scissors that allow you to snip genes out of an organism's DNA.

Gene editing can be used to remove genes that make an organism less nutritious or more **vulnerable** to disease. A recent example is the production of soybeans with more nutritious oil. The company Calyxt has produced a soybean seed with high oleic oil, which has healthier fats. This results from editing the DNA of the soybean rather than inserting genes from another organism.

No foreign genes have been added, so it is not transgenic. This process is not very different from older methods of using radiation on organisms until beneficial **mutations** emerge. It is just much faster, and the results are more predictable. As a result, there is some debate about how to label foods created through gene editing. Do they need the same labeling and testing as other GMOs to be considered safe to eat?

Soybean oil is made from soybeans. It is used as an ingredient in many processed foods.

The beta-carotene in golden rice gives it a deep yellow color. Developing this GE food is a project of the International Rice Research Institute.

Some GE seeds have functions not connected to their survival. GMOs can also have traits that make them more nutritious. Golden rice is a GMO still in development. The rice is modified with genes from daffodils and bacteria that provide a code for producing beta-carotene, a valuable nutrient. When we eat foods containing beta-carotene, like squash or carrots, it turns into vitamin A in our bodies. People who are vitamin A deficient do not have enough of the vitamin, and that can lead to blindness, illness, and death. Golden rice has been formulated as a staple food to provide beta-carotene for people in South and Southeast Asia, where vitamin A deficiency is most common.

Food for Thought

1. What are the advantages and disadvantages of using pesticides?
2. Are some genetic modifications better than others? Try comparing the effects of insect-resistant and herbicide-tolerant GMOs.
3. Should gene-edited foods be considered GMOs?

Monarch butterflies are one species affected by the increased use of herbicides related to HT crops. Glyphosate kills milkweed plants that monarchs rely on for survival.

Chapter Three

GMOs and the Environment

GMOs have a great potential to improve the resilience of crops and increase the availability of nutritious food in the world. However, not everyone is convinced that GMOs are a good thing. The past few decades have been an experiment in planting GE seeds, and the results have been mixed. Major environmental organizations like GreenPeace oppose the use of GMOs, shining a spotlight on the harmful effects GMO farming practices have had or could have on the environment. Concerns include reduced **biodiversity** and the loss of native species, the mixing of GMOs and wild species, and the increased use of chemical pesticides related to GMOs. Learning how to address these problems is essential to using GMOs wisely while protecting the environment.

Reduced Biodiversity

One of the biggest problems with GE crops is reduced biodiversity. The FAO reports that 75 percent of plant genetic diversity was lost between 1900 and 2000. Instead of planting local varieties of crops that are best adapted to their environments, farmers, encouraged by governments, have grown high-yield, genetically uniform crops. Farmers choose these crops because they are reliably productive. The problem is that if a disease or a pest were to attack, the whole planted crop—and consequently a large part of our food supply—would be affected because that whole crop has the same DNA.

The Irish potato blight is a famous example of the consequences of biodiversity loss. In the 1800s, farmers in Ireland were planting the same potato variety, the Irish Lumper, throughout the country. When an organism attacked these potatoes, it wiped out a major food source. The result was the

Growing multiple varieties of a fruit, such as a banana, can help prevent it from being completely wiped out by a disease.

Irish Potato Famine, which killed a million people and pushed a million more to emigrate.

A similar situation has occurred with bananas. Until the 1960s, the overwhelming majority of bananas eaten in the world were Gros Michel bananas. However, a fungus that causes Panama disease destroyed the crop. The banana that replaced it is called the Cavendish, and it accounts for about 47 percent of the bananas grown in the world. Now, over 50 years after its introduction as a banana resistant to Panama disease, it, too, is falling victim to a fungus called Tropical Race 4.

GMOs in the Wild

Another cause for concern is what's known as transgene escape or introgression, which is the idea that transgenic crops will cross with wild plants and create new wild plants with transgenes in their DNA. Some GE crops, like sunflower, canola, and sorghum, are at high risk of hybridizing, or breeding, with their wild relatives. A big fear is that the GE versions will pass herbicide tolerance to their wild relatives, which would make those relatives harder to kill. As a result, the GE crop would be harder to grow since the wild relative is unwanted and acts like a weed. Careful testing and consideration of this type of risk should be carried out with every GE product.

Building Resistance

Another issue with regard to weeds is herbicide tolerance caused by the use of HT crops. Because so many more farmers are spraying their crops with glyphosate, weeds that are not killed by the herbicide become resistant to it over time in order to survive. Glyphosate-resistant weeds have been found around the world, and there are about 38 weeds that have developed resistance since GMOs became popular. This

GMO Animals

Most GMOs developed in the last few decades have been plants. The first GE animal developed by scientists and approved as food by the FDA is a salmon called AquAdvantage. The DNA of the Atlantic salmon has been modified to produce this fast-growing fish. The AquAdvantage salmon takes less time than wild or other farm-raised salmon to grow to the right size for harvesting. In 2015, the FDA determined these fish have similar nutritional value to non-GE salmon and are safe to eat.

To prevent GE salmon from breeding with wild salmon, AquAdvantage fish can only be raised in tanks on fish farms. Containment rules were established in response to worries about the environmental impact of GE salmon if they escaped into the wild. Opponents of the GE salmon continue to voice their concerns about preventing introgression and preserving the rights, livelihood, and sacred traditions of Indigenous fishing communities like the Quinault Indian Nation in Washington State.

A genetically engineered pig was approved by the FDA in 2020. The GalSafe pig is free from a type of sugar called alpha-gal that many people are allergic to. The goal is to make food and medicines with pig products that are safe for people with this allergy.

AquAdvantage salmon grow to 8 to 12 pounds (3.6 to 5.4 kg) in 18 months. Atlantic salmon, like the fish pictured here, take 36 months to reach that size.

resistance requires farmers to spray more herbicides, thus creating a vicious cycle. Instead of reducing it, planting GE crops has actually increased the use of herbicides.

Insect resistance is a similar issue. Insects that are *Bt* tolerant survive, and those that aren't die. Over time, the *Bt*-tolerant trait gets passed down to future generations. To avoid this issue, the EPA requires farmers to plant refuges, which are areas next to GMO fields that are planted with non-GE varieties of a crop. Refuges allow insects that are *Bt* resistant to mate with non-resistant insects, decreasing the likelihood of resistant offspring.

Farmers with herbicide-resistant crops use chemicals to control weeds instead of using non-toxic weed management methods like crop rotation.

Patents and Pesticides

Genetically engineered corn, cotton, and soybean seeds have been developed and sold exclusively by the companies that produce them. Just six giant agricultural companies have controlled more than half of the world's seeds for many years. Monsanto and Bayer, Dow Chemical and Dupont, and China National Chemical and Syngenta merged in 2017 and 2018 to form even larger corporations.

These companies have patented their GE seeds. Patents are ways of protecting new ideas, also known as intellectual property. The idea is that thoughts are also property, and just as it is illegal to steal someone's property, it is illegal to steal their ideas. Seed corporations claim that patents for GE seeds are necessary to make back the money they invested into researching and developing the seeds.

As a result, when farmers buy seeds from these companies, they sign an agreement not to replant the seeds. This is a practice that goes back to 1930. Farmers have the option to save and reuse seeds that are not patented, but many choose not to because it takes a lot of time and money to clean and store those seeds for the next growing season. Furthermore, buying new seeds makes it more likely the plants will grow the way the farmer wants them to.

There are other staple crops developed by organizations other than corporations. GE wheat, for example, is largely developed by universities in the United States. Golden rice was developed by a nonprofit organization. Rainbow papaya was developed by academic and government researchers. Chemical companies play big roles in developing GE corn, soybeans, and cotton, but their research does not extend to many other staple crops, such as cassava, potatoes, and rice.

Although seed patents are not new, some people worry about letting one or two companies hold most of the existing patents.

Food for Thought

1. What are the benefits of biodiversity?
2. Is genetically engineering animals an ethical issue?
3. How do seed patents affect farmers? What does it mean to own a seed?

The Indigo Rose tomato is an example of a GMO that is engineered for greater nutrition instead of pesticide resistance. This tomato has extra nutrients that boost the health benefits of eating it.

Chapter Four

The Future of GMOs

Many of the environmental problems related to GMOs are the result of their use in a large system of industrial agriculture that relies on harmful practices like monocropping and dangerous chemical substances like pesticides. Some people would rather support **sustainable** farming practices, and so they may choose to avoid GMOs. Other people have ethical or health concerns that make them cautious about eating GM food. In order to respect the different viewpoints and values involved in this issue, it's important that foods with GMs are clearly labeled. Over time, bioengineering will continue to develop and more knowledge will be gained about GMOs and their impact on people and the planet. With these tools and knowledge, GMOs may play a major role in building a sustainable food system that produces enough nutritious food for all the people of the world.

Monocropping

Monocropping has become more popular with the introduction of GE crops. This is the practice of planting a single crop in

GMO Labels

When choosing foods for a healthy diet, it's important to know what's in them. Most foods in grocery stores come with labels that list ingredients and measure nutrients. Many people believe that GMOs should also be labeled. About 64 countries in the world have laws requiring GMO labels. People may choose to avoid GMOs because of environmental concerns, or because they worry that the long-term effects of using GMOs can't be measured yet.

There are a few ways to tell if packaged foods contain GMOs. First, look for a label that says Non-GMO. The Non-GMO Project was founded in 2007 to make it easier for people to buy non-GMO foods. Food producers can choose to have their food certified and marked with the organization's logo. However, just being non-GMO is not a guarantee that food was produced without pesticides. If you're looking for foods free from GMOs and pesticides, the best choice is a product that's certified organic by the USDA.

In 2016, Congress passed the National Bioengineered Food Disclosure Law. This law took effect in 2022 and requires that most foods with detectable GMOs are labeled with a logo, with the words "bioengineered" or "made with bioengineered ingredients," or with a phone number or code that provides facts about GMOs.

In the United States, non-GMO and organic labels give you information about how a certain food was produced.

the same field every year. Monocropping has allowed farmers to increase their yields without expanding the land they use. When farmers plant only one crop, a monoculture, they do not have to adjust the depth at which they plant the seed, they can use the same pesticide for the whole field, they can harvest more easily as the crops grow to a uniform height, and they don't have to sort through the harvest to separate out different products. Basically, monoculture allows farmers to mechanize the farming process, from planting to harvest. Mechanization uses more machines and fewer workers.

Monocropping would not be possible without the use of synthetic fertilizers. They increase productivity by providing plants with nutrients that promote growth, the most important of which are nitrogen, phosphorus, and potassium. Monocropping makes farmers very dependent on fertilizers since planting one crop repeatedly on the same land depletes that land of nutrients.

Not only does monocropping force an overreliance on synthetic fertilizer, but it also increases risks from pests and diseases. If a pest finds a crop it likes to feed on, it will stay in that area. To fight the pests that monocultures introduce, farmers use more pesticides. Monocropping can also increase disease since a disease can spread from one plant to another, so farmers have to rely on disease-resistant plant varieties. Disease resistance does not last forever, and pesticides become less effective with time. Both pesticides and synthetic fertilizers are harmful for the environment and for people.

Changing the crops in a field every couple of years, a process called crop rotation, is an alternative to monocropping and using synthetic fertilizers. It allows new crops to deposit nutrients in the soil that the previous crops removed. So, for example, while corn uses nitrogen from the soil, soy deposits

Bt cotton is farmed to make fabric, cottonseed oil for packaged and fried foods, and feed for animals. The overuse of Bt cotton means that some insects, like the cotton bollworm, have become resistant to Bt.

nitrogen in the soil. Rotating the two crops helps replace the nutrients that the crops consume. Crop rotation can also help control the effects of pests and diseases. Another option for using less fertilizer is polyculture—planting multiple crops at the same time. Polyculture, however, demands more workers in the field.

Working Toward Sustainability

GE seeds are not necessarily harmful to our health or the environment by themselves. However, the largest GE crops have been part of an agricultural system that is unsustainable. Another issue is the type of food produced. So far, soy and corn have been the most widely planted GE crops. These crops become ingredients in highly processed foods: corn, in the form of high-fructose corn syrup, and both corn and soy in the food fed to the animals that produce the processed meat sold by fast food restaurants (among others). These foods are directly linked to obesity and diseases related to diet, such as diabetes, high blood pressure, and some cancers.

What solutions are there, then, to the issues of modern industrial agriculture? It seems clear that monocropping is not a practice we can rely on. Crop rotation in and of itself would reduce the use of pesticides and synthetic fertilizers, and it would decrease the development of pesticide-resistant organisms. Polyculture, while beneficial for the soil, does not necessarily create the same yields as monoculture. It is also important to rethink the crops that we are planting. Is it necessary for the United States to grow as much corn as it does when so much of it—either directly or indirectly—goes toward producing food that is bad for us?

Because GE crops are not risky in and of themselves, they can contribute, along with other practices, to resolving

Over 70 percent of highly processed foods, such as packaged pizza, chips, cookies, and ice cream, are made with GMOs.

a fundamental challenge for humanity: feeding a growing population sustainably so that we do not use up the limited resources we have on Earth. GE seeds can help us improve nutrition, and they can allow us to grow food in a changing climate with more extreme droughts, floods, and heatwaves. A focus for the future can be on practices that won't deplete or pollute natural resources. We cannot continue to rely on pesticides and synthetic fertilizers to grow our food. Instead, the goal must be to use GE foods within a sustainable system of agriculture.

Food for Thought

1. Do you think people have a right to know what's in their food?
2. Under the National Bioengineered Food Disclosure Law, only packaged GMO food must be labeled as bioengineered. Do you think restaurants should also notify customers if their food has GE ingredients?
3. Do GMOs have a role to play in a sustainable food system?

Glossary

artificial selection: The process of deliberately crossing plants or animals to develop desirable characteristics, or traits, in future generations.

biodiversity: The variety of life in a particular ecosystem, habitat, or the world at large.

biophysicist: A scientist who uses physics to understand biological processes and structures.

crop rotation: The practice of changing planted crops year after year.

crossbreeding: Breeding an organism with an organism from another species, breed, or variety.

fertilizer: A material used on soil or on plants to provide plants with necessary nutrients for growth, including nitrogen, phosphorus, and potassium.

genetically engineered (GE): Refers to an organism whose DNA has been changed with the addition of DNA from another organism.

mutation: A change in the genetic makeup of an organism.

pesticide: A substance that prevents, repels, or destroys pests, which can come in the form of insects, weeds, fungi, and rodents.

processed food: Food that has been significantly changed from its original form through chemical or mechanical methods.

resilient: Able to withstand challenges.

resistance: The natural ability of an organism to survive or avoid a substance that is toxic to it.

sustainable: Involving processes that don't destroy or completely use up the planet's natural resources.

vitamin: Any of a number of nutrients, found in fruits, vegetables, grains, fungi, meats, and dairy foods, essential to the human body for health.

vulnerable: Likely to be harmed.

Find Out More

Books

Mason, Paul. *Making Our Food Sustainable*. New York, NY: Crabtree Publishing Company, 2019.

Mooney, Carla, and Traci Van Wagoner. *The Chemistry of Food*. Ashland, VT: Nomad Press, 2021.

Woollard, Alison, and Sophie Gilbert. *The DNA Book*. New York, NY: DK Publishing, 2020.

Websites

Agricultural Marketing Service
www.ams.usda.gov/rules-regulations/be/bioengineered-foods-list
This list of bioengineered foods available for sale in the United States is updated regularly by the USDA's Agricultural Marketing Service.

National Geographic
www.nationalgeographic.org/article/are-genetically-modified-crops-answer-world-hunger/
The article "Are Genetically Modified Crops the Answer to World Hunger?" explores the potential benefits of GMOs for feeding the world.

World Health Organization
www.who.int/news-room/questions-and-answers/item/food-genetically-modified
This question and answer page from the WHO provides a wealth of information about GMOs and their effects on health and the environment.

Organizations

Feed the Future
1300 Pennsylvania Avenue
Washington, DC 20004
feedthefuture@usaid.gov
www.feedthefuture.gov
Feed the Future is a government organization in the United States that addresses world hunger by collaborating with other countries and organizations to develop innovative food solutions.

International Rice Research Institute (IRRI)
10th Floor Suite 1009
Security Bank Center
6776 Ayala Avenue
Makati, Metro Manila 1226
Philippines
www.irri.org/
The IRRI is a nonprofit group that has developed various strains of genetically modified rice to end hunger and prevent diseases related to malnutrition.

The Non-GMO Project
1155 North State Street
Bellingham, WA 98225
www.nongmoproject.org/
The Non-GMO Project is a nonprofit organization that aims to help people identify foods made with GMOs and advocate for clear labels.

Publisher's note to educators and parents: Our editors have carefully reviewed these websites to ensure that they are suitable for students. Many websites change frequently, however, and we cannot guarantee that a site's future contents will continue to meet our high standards of quality and educational value. Be advised that students should be closely supervised whenever they access the internet.

Index

About the Author

Rachael Morlock is a freelance writer and copyeditor. She is the author of several nonfiction and picture books for children and enjoys researching new subjects through her work. In the past, she worked as the coordinator of a Community Supported Bakery. Rachael is always interested in supporting local farmers and food producers in Western New York, where she lives with her dog, Rilke.